Foundation for Neurofeedback
& Neuromodulation Research

## FNNR Publications

Sokhadze, E., Morosoli, T., Frederick, J., Evans, J. (2023). *International Trends in Neurofeedback Volume 1.* Foundation for Neurofeedback and Neuromodulation Research.

Longo, R. E., & Bingham, B. (2022). *Beginning Neurofeedback in Your Practice: From Intake to Discharge.* Foundation for Neurofeedback and Neuromodulation Research.

Soutar, R., & Longo, R. E. (2022). Doing Neurofeedback: An Introduction (2nd Ed.). Foundation for Neurofeedback and Neuromodulation Research

Bester, Helena (2020). Neurofeedback: The Non-invasive Alternative. Foundation for Neurofeedback and Neuromodulation Research.

Longo, R. E., & Soutar, R. (2020). *Becoming Certified in Neurofeedback: A Guide to the Neurofeedback Mentoring Process for Mentors and Mentees.* Foundation for Neurofeedback and Neuromodulation Research.

Sokhadze, E., & Casanova, M. F. (Eds.). (2019). *Autism Spectrum Disorder: Neuromodulation, Neurofeedback, and Sensory Integration Approaches to Research and Treatment.* Foundation for Neurofeedback and Neuromodulation Research.

Martins-Mourao, A., & Kerson, C. (Eds.). (2017). *Alpha-Theta Neurofeedback in the 21st Century: A Handbook for Clinicians and Researchers* (2nd ed.). Foundation for Neurofeedback and Neuromodulation Research.

Soutar, R., & Longo, R. E. (2011). *Doing Neurofeedback: An Introduction.* The ISNR Research Foundation.

Hammond, D. C., & Gunkelman, J. (2011). *The Art of Artifacting.* The ISNR Research Foundation.

Carmichael, J. (2011). *Multi-Component Treatment for Post-Traumatic Stress Disorder, Including Strategies From Clinical Psycho-Physiology and Applied Neuroscience.* The ISNR Research Foundation.

Donaldson, S. (2012). *The Other Side of the Desk.* The ISNR Research Foundation.

# RICHARD SOUTAR, Ph.D., BCN

## &

# ROBERT LONGO, MRC, LCMHC, NCC, BCN

Publisher:

The FNNR

The Foundation for Neurofeedback &
Neuromodulation Research

101 Stewart Street Hendersonville, NC 28792

http://www.theFNNR.org

Correspondence: admin@theFNNR.org

Publisher: The FNNR ,The Foundation for Neurofeedback & Neuromodulation Research
101 Stewart Street Hendersonville, NC 28792
http://www.theFNNR.org. Correspondence: Admin@theFNNR.org
Layout Design: Shea Connor. Correspondence: publishing@theFNNR.org
Cover by: Deb Longo
Distributed by: BMED Press, LLC
FUNCTIONAL NEUROFEEDBACK: AN INTRODUCTION

ISBN: 979-8-9889374-1-8

# Table of Contents

# ACKNOWLEDGMENTS

I would like to begin back thanking my colleague and co-author Dr. Richard Soutar. Richard was my initial mentor when I began learning about neurofeedback and has always encouraged me to learn and move forward with providing neurofeedback services.

Over the course of the past 17 years, we have worked together to help educate colleagues and other practitioners about neurofeedback training, QEEG brain mapping, QEEG interpretation and findings, and practicing neurofeedback from a holistic, bio-psycho-social model. These concepts have evolved into the functional neurofeedback model which Richard coined and has developed over the past few years.

I would like to also acknowledge the clients and patients I have worked with over the years who have taken the time to give me detailed histories, completed numerous forms and assessments, and took the time to fill out weekly progress reports that outlines the reduction of symptoms and benefits of participating in neurofeedback.

In addition, I'd like to acknowledge and thank my colleagues Judy Prichason and Veronica Seberger who participate as panelists in the weekly webinars we conduct and provide ongoing education, resources, and support for this model.

Finally, I'd like to thank my wife Deb Longo who has always been supportive of the time I take to write and co-lead weekly webinars.

# FOREWORD

*It is a pleasure to contribute a foreword to this new text, especially given its presentation of the vision and practice of functional neurofeedback (FNFB). The authors begin with an excellent presentation of the growing contributions from the field of functional and integrative medicine—which is in many ways, a return to the heart of true healthcare— evaluation of the whole person in order to promote disease amelioration, and restoration of health. The superficial practice of symptom-identification and symptom-treatment may offer temporal relief on occasion (such as treating a fever in a person with meningitis), but does not address the underlying causes. The FNFB approach is a comprehensive healthcare model which goes far beyond "disease management" to strive for optimal wellness through a comprehensive process.*

*Each chapter builds on the previous one. After laying out the foundation of FNFB (chapters 1–2), they discuss integrating the Holistic Health Model (chapter 3), naturally following with outlining the processes involved in diligently formulating a FNFB training plan—while working within one's scope-of-practice. What follows is practical wisdom and direction in the "how-to" implement this whole-person approach in one's clinical practice (chapter 5), concluding with a realistic overview of the challenges that the clinician faces while integrating FNFB into a comprehensive, whole-person paradigm of practice. This approach involves much from the clinician—requiring significant planning and time investment to help one's client/patient to look at all aspects of their life, come to an increasing awareness of the many factors contributing to their concerns/symptoms, and offering practical solutions toward restoring health and wellness.*

-Robert P. Turner, MD, MSCR, QEEGD, BCN

# INTRODUCTION

To best understand functional neurofeedback, it is important to understand the development and purpose of functional medicine. According to the Institute for Functional Medicine (2024), "Functional medicine determines how and why illness occurs and restores health by addressing the root causes of disease for each individual."

"The functional medicine model is an individualized, patient-centered, science-based approach that empowers patients and practitioners to work together to address the underlying causes of disease and promote optimal wellness. It requires a detailed understanding of each patient's genetic, biochemical, and lifestyle factors and leverages that data to direct personalized treatment plans that lead to improved patient outcomes.

By addressing root cause, rather than symptoms, practitioners become oriented to identifying the complexity of disease. They may find one condition has many different causes and, likewise, one cause may result in many different conditions. As a result, functional medicine treatment targets the specific manifestations of disease in each individual (Institute for Functional Medicine, 2024).

In December of 2021 Richard Soutar coined the term functional neurofeedback while describing a holistic model of neurofeedback service delivery (Soutar, 2021). The term was so relevant to the process he and NewMind Technologies use to evaluate, assess, and deliver neurofeedback services, we decided to define and enhance this model and method of providing neurofeedback services to clients and patients. In

November 2023, NewMind Technologies released its functional neurofeedback training software.

In 2018, Richard Soutar released his monograph titled: Holistic Neurointegration: The NewMind Model - A Bio-Psycho-Social QEEG Guided Neurofeedback Method. This monograph, an evolving E-book, describes the NewMind model and incorporates the various aspects of what we call functional neurofeedback. This method evolved over the course of 15 years as he developed the NewMind mapping system which includes a comprehensive look at the client's/patient's history, background and lifestyle including diagnosis, medications, physical health history, personality traits, and cognitive and emotional assessments.

The idea of functional neurofeedback is new and not widely known or acknowledged to date. However, we propose the following definition:

Functional neurofeedback is an integrative model of care that offers a client/patient focused approach to addressing disorders of the brain and central nervous system. The goal is to provide personalized client/patient-centered care. We recognize that each symptom or differential diagnosis may be one of many contributing to an individual's illness.

Functional neurofeedback clinicians complete a comprehensive intake and assessment and take a comprehensive health history. The clinician uses this information to better pinpoint the root cause(s) of brain, central nervous system and mental health disorders that include but are not limited to diet and nutrition, physical health (metabolic and physiological), stress levels, environment factors, lifestyle, and sleep hygiene.

The goal of functional neurofeedback is to provide QEEG Brain Mapping for assessment purposes and brainwave training (neurofeedback) to address disorders of the brain and central nervous system in order to promote a healthy lifestyle. Not to just resolve a specific disorder

The foundational approach to functional neurofeedback includes:

- A comprehensive assessment including cognitive, emotional, and physical health evaluation.

- Supporting a comprehensive lifestyle with a focus on diet and nutrition, exercise, social relationships, and sleep hygiene.

- Patient-centered care using complimentary health care guidance

- Integrative, science-based interventions including but not limited to biofeedback, stress management, methodology, lifestyle changes.

- Multidisciplinary approach integrating best physical and mental health practices including appropriate and comprehensive assessment and testing, and prevention strategies.

# CHAPTER ONE

## What is Functional Neurofeedback?

The emergence of individualized medicine and evidence-based medicine has brought with it a new paradigm that has become known as functional medicine as envisioned by Jeffery Bland (Bland, 2018; 2019a). Functional medicine seeks to treat the source of the patients' problems rather than just treating the symptoms and consequently forces the practitioner to transcend the one size fits all approach of acute care medicine in its institutionalized form. Functional medicine is less concerned with the dysfunction of a disease and more concerned with the dynamic processes that generate disease. Disease is a descriptor of a pathological endpoint and function is about the sources of the pathological process and how it is shaped by multiple causal factors in a biological system. It is a scientific systems-based perspective (Bland, 2017; 2019b). The importance and validity of this rapidly developing approach has been recognized in JAMA (Chang & Lee, 2018). The Institute for Functional Medicine defines functional medicine as follows (IMF; https://www.ifm.org):

"Functional medicine determines how and why illness occurs and restores health by addressing the root causes of disease for each individual."

### *Some key features of functional medicine include:*

- It addresses root cause rather than just symptoms.
- It recognizes the complexity of contributing factors to disease.
- It promotes "optimal wellness."

- Assessment requires a detailed understanding of each patient's genetic, biochemical, and lifestyle factors.
- It is individualized.
- It is patient centered.
- Patient and practitioner work together.
- It empowers the patient.
- It is science based but not randomized controlled trial (RCT) dominated.
- It considers lifestyle a critical factor.

Functional medicine physicians do exhaustive patient interviews covering all aspects of their lives and then form hypotheses regarding the potential sources of the patient's ailments and symptomology. They then utilize scientific testing both in initial assessment as well as in a methodical and strategic manner with each intervention to assess its efficacy and help determine whether associated side-affects justify such an intervention. In this step-by-step fashion physicians eventually isolate the source of the patient's suffering and can effectively treat them. Functional neurofeedback seeks to emulate this model.

This approach stands in stark contrast to traditional acute care and chronic care institutions that have formalized diagnostic and treatment protocols that apply equally to any patient engaging the institution regardless of whether it is an ER, hospital or mental health facility. Many of these protocols involve standard medications and treatments that overlook individual differences in response and efficacy. Functional neurofeedback at the same time stands in stark contrast to other neurofeedback approaches that seek to "fix" a symptom by training at one defined region of interest (ROI) supposedly associated with that symptom or the use of one protocol to treat one label such as beta up and theta down for all ADHD clients while disregarding family dynamics, nutrition, sleep and other contributing factors.

## The Functional Approach Concept and Neurofeedback (NFB)

Early in the field of neurofeedback there was a developing paradigm that encouraged practitioners to see NFB as a panacea that drew its strength from treating the CNS as the ultimate solution. Many in the field felt that by training the central nervous system, most human afflictions might be overcome. To some extent this was driven by unexpected and impressive outcomes in many inexplicable cases that seemed to border on the miraculous (Robbins, 2000). In response to these cases Jim Robbins, a science writer for the New York Times authored an article in Psychology Today that was entitled "Wired for Miracles" (1998).

An expectation emerged that with enough sophisticated technology all obstacles could be overcome. Much of the product development was driven by engineers and entrepreneurs who had little or no clinical experience and training. Based more on clinical experience than research, clinicians fell in line and this perspective matured and to some extent temporarily dominated the field. If a client failed to respond to treatment, then it was considered due to a failure of technology and such issues would be eventually resolved through future product development or enhanced techniques alone.

This perspective was paralleled in the field of mental health and the psychiatric community's effort to join the ranks of experimental deductive sciences by deploying drugs designed specifically to treat each specific and categorized disease (van der Kolk, 2014; Whitaker, 2010). If one drug did not work, then eventually RCT research would find a better drug that did the job. This approach found favor in the insurance companies seeking an effective and efficient way to manage health claims and pharmaceutical companies that gain massive sales through marketing the same concept. Unfortunately, mental health outcomes got worse rather than better and the DSM fell into disrepute (Healy, 2004; Whitaker, 2010).

In 2014 the Cleveland Clinic for Functional Medicine was established. Dr. Mark Hyman, author of 12 New York Times bestsellers, in conjunction with the Cleveland clinic and others, have been leading the change in medicine with the functional medicine paradigm with growing evidence of significantly better outcomes than previous approaches. The personalized lifestyle healthcare model is co-evolving with the functional medicine model (Bland, 2018). At the same time, they are finding this approach is resolving many mental health issues as well, supporting with evidence the theory that many disorders have a strong basis in physical health problems.

In both the mental and physical health paradigms a strong awareness has been growing those multiple vectors of causality are at work with health disorders. The same applies to disorders such as ADHD which is defined by multiple etiological pathways rather than being a unidimensional construct with one remedial protocol or solution (Núñez-Jaramillo et al., 2021). The DSM and psychiatric community have been very reluctant to recognize the significant contribution of social and ecological factors to poor health despite the findings of the Adverse Childhood Experiences (ACE) study, one of the largest and most comprehensive longitudinal health studies conducted to date, as well as position papers by the American Pediatric Association (van der Kolk, et al. 2016; Shonkoff & Garner, 2012). The research evidence is quite clear that mental and physical health problems emerge from a combination of social and ecological system contributions and that those findings engender a perspective that threatens established economic, political, and cultural paradigms as is commonly the case in institutionalized science (Kuhn, 1970).

New Mind technology was founded upon the multi-disciplinary research findings driving these same changes in medicine in general. We have promoted a bio-psych-social approach to neurofeedback from our inception that is in harmony with the functional medicine approach. Consequently, such a perspective may be characterized as functional neurofeedback. A key concept in this approach is that electronic technology and interventions by themselves will never be sufficient alone to treat disorder regardless of how many iterations it pursues. Coben et al. (2018) note that most of these highly evolved technical approaches to neurofeedback have not been demonstrated to be superior to traditional neurofeedback by any significant research to date. Truly efficient remediation of symptoms requires an inclusive awareness of the contributions of multiple causal factors arising from social and ecological sources.

The assessments constructed for the New Mind Database System were built in concert and from the ground up for the purposes of applying them from a psychophysiological perspective, rather than a medical or

15

psychiatric perspective, to a bio-psycho-social approach to neurofeedback. Consequently, they comprise a highly consistent and integrated assessment tool for tracking the effects of complex clinical protocols that define the functional approach. This was done in recognition of the many clinical confounds that have to be accounted for in successful clinical outcomes. The system was constructed using standard and universally recognized statistical methods with one of the largest and fastest growing clinical neurofeedback databases in existence with over two million subjects. The instruments cover traditional physiology, socio-emotional, psychological and cognitive dimensions from an integrative psychophysiological perspective that is more holistic and systems based in its approach than other academically and traditionally isolated efforts of understanding health and wellness.

New Mind enables practitioners to assess multiple dimensions of psychophysiological dysregulation and their global contributions to presenting symptoms including a physiological checklist, a socioemotional assessment, a cognitive emotional checklist, computerized cognitive performance tests and QEEG brain maps. This suite of statistically validated assessments provides a more comprehensive understanding of the client and the sources of dysregulation. It also provides the practitioner with a good foundation for working in concert with other professionals in nutrition and functional medicine as well as other specialists. By using multiple measures in conjunction with QEEG driven neurofeedback, a one size fits all approach to client wellness is avoided and a more individualized and empirical approach is encouraged. With the addition of neurofeedback, we have found outcomes significantly accelerated and enhanced and easy to document using our symptom tracker and pre-post brain mapping tool. In this manner, New Mind operationalizes the functional approach and fulfills in parallel fashion in terms of neurofeedback and psychophysiology the client centered individualized vision inherent in functional medicine as defined by Bland and his associates.

# CHAPTER TWO

## What's the Difference between Neurofeedback and Functional Neurofeedback?

The International Society for Neuromodulation and Research (ISNR; 2024) states, "Like other forms of biofeedback, neurofeedback training (NFT) uses monitoring devices to provide moment-to-moment information to an individual on the state of their physiological functioning. The characteristic that distinguishes NFT from other forms of biofeedback is a focus on the central nervous system and the brain. NFT has its foundations in basic and applied neuroscience as well as data-based clinical practice. It considers behavioral, cognitive, and subjective aspects as well as brain activity"

"NFT is preceded by an objective assessment of brain activity and psychological status. During training, sensors are placed on the scalp and then connected to sensitive electronics and computer software that detect, amplify, and record specific brain activity. Resulting information is fed back to the trainee virtually instantaneously with the conceptual understanding that changes in the feedback signal indicate whether or not the trainee's brain activity is within the designated range. Based on this feedback, various principles of learning, and practitioner guidance, changes in brain patterns occur and are associated with positive changes in physical, emotional, and cognitive states. Often the trainee is not consciously aware of the mechanisms by which such changes are accomplished although people routinely acquire a "felt sense" of these positive changes and often are able to access these states outside the feedback session" (ISNR, 2024).

The ISNR Bylaws state (ISNR, 2024): "To promote excellence in clinical practice, educational applications and research in applied neuroscience in order to better understand and enhance brain function."

***Our Objectives Are...***

1. Improve lives through neurofeedback and other brain regulation modalities

2. Encourage understanding of brain physiology and its impact on behavior

3. Promote scientific research and peer-reviewed publications

4. Provide information resources for the public and professionals

5. Develop clinical and ethical guidelines for the practice of applied neuroscience.

Functional neurofeedback (FNFB) takes into account all of the above and also holds that the human body is an entire system that includes physical health, mental health, and neurological/nervous system functioning. In an effort to maintain optimum health in these areas, FNFB includes the daily monitoring of diet and nutrition, exercise, social relationships, sleep hygiene, and the impact of environmental factors that negatively affect our health (i.e., blue light from electronic device screens and electro-magnetic fields; EMFs), to support and enhance optimum health. It is not a singular approach to working with a specific disorder.

According to the Institute for Functional Medicine (IFM), "The functional medicine community comprises a diverse array of clinicians interested in addressing the root causes of their patients' health issues to improve outcomes. One of the tenets of functional medicine is to understand each patient's story, the factors which may have predisposed that individual to a chronic condition, and what may have triggered the onset of the condition.

A key tool in this process is the functional medicine timeline. This intuitive tool provides a framework for tracking the patient's story over time. Frequently, when patients view their health journey on a timeline, they make new realizations about their health." (The Institute for Functional Medicine, 2024).

FNFB clinicians form a network of health care providers for purposes of referral for those clients/patients that require the supportive care of other health care professionals including but not limited to neurologists, integrative medicine doctors, functional medicine doctors, chiropractors, registered nutritionists, psychologists, social workers, counselors, and physical therapists.

"Some questions clinicians may want to consider prior to developing a plan for lifestyle change include:

**Is your patient ready to consider this change?**

What might your patient need to give up in order to embrace change?

What barriers might your patient face that might discourage lifestyle change?

**How has your patient successfully changed behavior in the past?**

A health provider can play a critical role in this process by increasing patient awareness through a collaborative approach to sharing information, education, and personal feedback. The functional medicine model was developed, in part, to do just this. Functional medicine is grounded in a strong patient/provider relationship. Specific tools—like the IFM Timeline and Matrix (The Institute for Functional Medicine, 2024)—help the practitioner understand the course of the patient's life as seen through the lens of health and disease. Often, disease occurs when fundamental lifestyle factors like diet, movement, rest, and/or sleep are lacking or imbalanced in an individual's life."

**Diagnostics and Labels**

Diagnostic labels are generally based upon symptoms and not causes. This method of labeling clients/patients with disorders is not always effective in creating a comprehensive treatment plan. Traumatic brain injuries (TBI) are a good example. TBI symptoms often mimic the following disorders: ADD, ADHD, anxiety, depression, PTSD, bipolar disorder, learning disability, and oppositional defiance in youth to name a few (Swatzyna, 2009).

The list of symptoms that can occur from head injuries and traumatic brain injuries is extensive. They include but are not limited to:

Agitation
Aggression
Alcohol use (increased)
Altered executive function
Anxiety (panic attacks)
Apathy
Ataxia (unsteadiness)
Attentional problems
Balance problems
Blurred vision
Change in sense of smell
Change in vision
Change in menses (periods)
Coldness
Confusion
Decrease in intelligence
Decrease in recent memory
Decrease in remote memory
Depression
Dizzy (vertigo)
Ear infections
Emotional lability
Fatigue
Focusing problems
Foggy headed
General intelligence deficits
Headache
Hearing deficits
Impulsivity
Impaired judgment
Increase in tiredness or fatigued
Information processing
    problems/slowed processing

Irritable/angry
Lack of concentration (focus)
Lack of interest in life/ bored
Lack of sex drive (libido)
Lack of competitiveness
Lack of confidence
Language problems
Lightheadedness
Memory problems
Moody/mood swings
Motor skills deficits
Nausea
Obsessive thoughts
Onset of insomnia
Periods of disorientation
Poor memory
Recurrent headaches/
    migraines
Seizures
Self-isolation
Sensitivity to light/noise
Sinus problems
Sleep problems
Sleeping more
    (hypersomnia)
"Spacey" spaced out
Substance abuse
Sudden outbursts of Anger
Sudden irritability
Walking/gait problems
Weakness
Weight gain
Weight loss

When reviewing the list of symptoms above, how many diagnoses have three or more of those in the above list?

When the DSM-V was published in 2013, it was immediately challenged by then NIMH Director Tomas Insel MD. "The DSM—the so-called bible of mental health professionals—has long been the primary tool for diagnosing and classifying mental illnesses. But in a blog post, NIMH Director Thomas Insel, MD, said that it was time to "re-orient" away from the DSM's symptom-based categories. Instead, the institute is developing a new framework called Research Domain Criteria (RDoC) that replaces the DSM diagnoses with broader research categories that incorporate behavioral and neuroscience evidence." (Winerman, 2013; Lane, 2013).

"In order to avoid alienating any particular constituency of mental health professionals, the DSM has strategically adopted an atheoretical stance on the etiology or causes of mental disorders in its definitions. At the same time, the DSM conforms to a medical model by organizing mental disorders into discrete categories, just as medicine does with diseases. That is, the DSM is a medical-model manual that is nonetheless atheoretical about the causes of the mental disorders it catalogs. This may be confusing but important to keep in mind." (Raskin, 2012).

In her talk on "The Traumatized Brain: Recent Neuroscience Findings," Sebern Fisher (2022), notes that the only diagnostic category in the DSM-V that provides possible etiology/causes is PTSD.

Within the field of neurofeedback, several researchers have begun to reject the diagnosis of ADHD as not a "correct" diagnostic category, and that in fact, most ADHD is the result of significant stress, anxiety, and poor sleep (Krepel et al., 2020).

A study, published in Psychiatry Research (Science Daily, 2019), has concluded that psychiatric diagnoses are scientifically worthless as tools to identify discrete mental health disorders. The study, led by researchers from the University of Liverpool, involved a detailed analysis of five key chapters of the latest edition of the widely used Diagnostic and Statistical Manual (DSM), on "schizophrenia," "bipolar disorder," "depressive disorders," "anxiety disorders" and "trauma-related disorders" (Allsopp et al., 2019).

The main findings of the research were (Allsopp et al., 2019):

- Psychiatric diagnoses all use different decision-making rules.
- There is a huge amount of overlap in symptoms between diagnoses.
- Almost all diagnoses mask the role of trauma and adverse events.
- Diagnoses tell us little about the individual patient and what treatment they need.

The authors conclude that diagnostic labelling represents "a disingenuous categorical system." In other words, psychiatric diagnosis is "scientifically meaningless."

Kaiser Permanente in conjunction with the Centers for Disease Control have been conducting the Adverse Childhood Experiences Study (ACEs) began in 1994 and includes over 17,000 adults. Adverse childhood experiences (ACEs) can have a tremendous impact on future violence victimization and perpetration, and lifelong health and opportunity.

"ACEs are linked to chronic health problems, mental illness, and substance use problems in adulthood. ACEs are common. About 61% of adults surveyed across 25 states reported that they had experienced at least one type of ACE, and nearly 1 in 6 reported they had experienced four or more types of ACEs."

The ACES study demonstrates immune systems are compromised. When immune systems are compromised it can affect the entire human body including the brain. Inflammation often shows up in quantitative electroencephalography (QEEG).

Functional neurofeedback looks well beyond symptoms, diagnostics, and single interventions, and one size fits all health care treatment plans.

## Marketing

When a clinician offers NFB services it is important how they market their services for several reasons (Longo & Bingham, 2022). Incorrect marketing can result in ethical violations leading to possible licensure concerns due to scope of practice violations (Longo & Sherlin, 2022). Thus, another difference between NFB and FNFB is how practitioners market their services.

In recent times some neurofeedback practitioners have come under fire for how they market their services. Wexler et al. (2019) in their paper titled, "Neuroenhancement for Sale: Assessing the Website Claims of Neurofeedback Providers in the USA" note, "In sum, there is a considerable divergence between the scientific literature on neurofeedback and the marketing of neurofeedback services to the general public, raising concerns regarding the misrepresentation of services and misleading advertising claims."

For example, a recent article in Neuroscience News, "How Not to Use Brain Scans in Neuroscience" (France, 2022), summarizes the research stating, "While neuroimaging may be a standard in neuroscience and psychology research, a new study says researchers are massively underestimating how large the study sample must be for a neuroimaging study to produce reliable findings."

In the article, Brenden Tervo-Clemmens, a postdoctoral fellow at Massachusetts General Hospital and Harvard Medical School who co-led the multi-institutional research as a clinical psychology PhD student at Pitt notes, "We can count on less than a hand the number of these studies that have held up under scrutiny and are really driving treatment," he said. "In my own area, one study might show that increased function of a particular brain region is related to more symptoms, but you can find, almost without question, another study showing the opposite effect."

**Branding**

Longo & Bingham (2022) state, "Branding is the process of building awareness, loyalty and engagement, which brings clients to your door and awareness to your community. It allows your business to stand out and be distinct from other similar businesses or from other alternative choices. It involves marketing, advertising, and selling via many different avenues."

Before you dive into branding and marketing, you need to consider the audience you'll serve, the disorders you'll specialize in, and the scope of your practice based upon your degree and licensure.

## Scope of Practice

When focusing on functional neurofeedback, scope of practice must be taken into account. Why? Because functional neurofeedback providers work with the whole person and the entire human health system; and medical disorders must be diagnosed and treated by a licensed medical doctor or other licensed healthcare provider. The FNFB provider works in conjunction with those health care professionals in a collegial fashion such that the diagnosing health care provider is made aware of and supportive of FNFB as an intervention, i.e., working with medical disorders such as seizures, TBIs, migraine headaches, insomnia, ADHD, etc.

Longo and Bingham (2022) and Longo and Sherlin (2022) note, "Scope of practice is one of the areas in which a practitioner can easily get him or herself into trouble. One should make sure that the disorder you are working on is within your scope of practice or work with a specialist who can provide you with clinical guidance or supervision for a particular disorder. Neurofeedback clinicians may elect to address the symptoms of anxiety, depression, relaxation, and sleep problems. Therapists who have a particular specialty such as addictions may focus their neurofeedback training on this particular group of clients."

If you plan to limit your practice to specific disorders, then it is best to note your specialty areas in your marketing materials. We encourage practitioners to use the term "brain wave training" versus "neurofeedback treatment" to avoid stepping outside your scope of practice. "Training down negative symptoms" is one way to phrase the use of neurofeedback versus saying that one is "treating a patient for ADHD," which is a diagnosable medical disorder. As seasoned neurofeedback clinicians we have seen some of our colleagues have their practices closed because they were "treating" a medical disorder and thus charged with practicing medicine without a license.

Marketing your neurofeedback services is extremely important. There are many different avenues you can use to market your services. Neurofeedback is not a common service, like roofing, plumbing, or electrical work, where you can often merely advertise the name of your service and people immediately understand what you do. Neurofeedback marketing often requires a strong educational component. This excludes some marketing avenues that rely upon immediate category recognition and do not offer space for educational content, such as banner ads, school program ads, etc. We will describe some of the most valuable marketing avenues that allow an educational component here. "

In general, neurofeedback practitioners work with patients who have a particular symptom/set of symptoms and/or specific disorder(s). Symptoms are targeted, often through the use of a QEEG and then protocols used based upon the QEEGs findings to target the problematic symptom identified by the client/patient, i.e., ADD. The FNFB practitioner looks at the reported problem/diagnosis (i.e., ADD) and works to identify the causes of the diagnosis/disorder, for example a child diagnosed with ADD but a history of stress, anxiety, poor sleep and poor diet.

# CHAPTER THREE

## A Holistic Health Model

Historically, learning about neurofeedback focused on the human brain. However, over the past decade the scientific community has repeatedly and consistently shown connections between the brain and other body systems. The most popular connection in recent years has been the "gut-brain" connection.

Carabotti et al. (2015) note, "This interaction between microbiota and GBA appears to be bidirectional, namely through signaling from gut-microbiota to brain and from brain to gut-microbiota by means of neural, endocrine, immune, and humoral links."

Other connections have also been addressed in science literature. For example, vitamin B-12 deficiency can lead to anemia. A mild deficiency may cause no symptoms. But if untreated, it may lead to symptoms such as:

Weakness
Tiredness
Lightheadedness
Heart palpitations
Shortness of breath
Pale skin
A smooth tongue
Constipation
Diarrhea
Loss of appetite
Flatulence
Nerve problems
       Numbness
       Tingling
       Muscle weakness

> Problems walking

Vision loss

Mental health
> Depression
> Memory loss
> Behavioral changes

Black (2011) reports that folate deficiency in the periconceptional period contributes to neural tube defects; deficits in vitamin B12 (cobalamin) have negative consequences on the developing brain during infancy; and deficits of both vitamins are associated with a greater risk of depression during adulthood.

Dholakia et al. (2005) report that vitamin B-12 deficiency can lead to GI symptoms and severe vitamin B-12 deficiency can cause a range of gastrointestinal symptoms including gastritis and atrophy.

These brief examples are just partial explanations of how a comprehensive focus on health can provide the clinician with invaluable information regarding an individual's overall health and well-being.

We propose several health-related avenues clinicians should take in conducting an intake and assessment of perspective NFB clients/patients. A comprehensive evaluation should include demographics, family history, social history, a history of any abuse or neglect, trauma history, comprehensive physical health history, comprehensive mental health history, lifestyle including diet and nutrition, exercise, socialization, work, hobbies, use of electronics and social media, sleep hygiene, etc.

From a physical health perspective, we pay close attention to physiological and metabolic categories. Some of the metabolic categories we gather information on include:

Adrenals
Blood sugar
Cardiovascular
Gall bladder
Gastrointestinal
Kidney
Liver
Pituitary
Somatic
Thyroid (hypo)
Thyroid (hyper)

These are all important areas of concern because they can seriously affect health and in turn effect brain function. The above, and other areas addressed earlier are some of the many reasons we are promoting a holistic health model.

**What is Holistic Health?**

The Institute for Holistic Health Studies at Western Connecticut State University defines holistic health as: "Holistic health is an approach to life that considers multidimensional aspects of wellness. It encourages individuals to recognize the whole person: physical, mental, emotional, social, intellectual, and spiritual. The individual is an active participant in their health decisions and healing processes, including wellness-oriented lifestyle choices. Holistic approaches to health are derived from ancient healing traditions that help to achieve higher levels of wellness and prevent disease. These approaches include use of traditional medical systems, mind-body-spirit interventions, manipulative and body-based approaches, biological based therapies, and energy therapies. Most of these approaches are used in combination with each other and with conventional medicine to provide a holistic and integrated approach to health. These traditional holistic approaches focus on the use of food, herbs, supplements, teas, homeopathic remedies, and essential oils as "medicine." Movement, dancing, singing or chanting, sound and vibration, drumming, prayer, meditation, mindfulness, and touch are examples of activities that are included in holistic approaches. Holistic approaches include but are not limited to: acupuncture, acupressure, biofeedback, massage therapy, chiropractic physicians, manual therapy, naturopathic physicians, meditation, guided imagery, yoga, therapeutic touch, reiki and other energy therapies, and ayurveda. On campus, the interest and enthusiasm for this inclusive and multidimensional approach to health and wellness has resulted in the development of a concentration in Holistic and Integrative health within the Health Promotion Studies major at WCSU."

Our belief is that a holistic health/bio-psycho-social model of health care fits perfectly into the functional neurofeedback paradigm. When FNFB includes this paradigm, comprehensive treatment planning is enhanced, and optimum care occurs.

Robert "Rusty" Turner, MD, our friend, colleague, and owner of Network Neurology Health is a neurologist who has developed what he calls the M.E.D.S. model.

Simply stated, M.E.D.S. stands for:

**M**inimize: Screen time, exposure to blue-light & EMF effects. Disconnect digitally. Healthy screen, media, and electromagnetic device use.

Exercise: Move and exercise regularly.

Diet & Nutrition: Drink water, eat healthier and healthier.

Sleep: Healthier and healthier. For optimal health: SLEEP, REST, RELAX, and BREATHE.

AND… follow the Golden Rule: Wash your hands with soap and water frequently and be always respectful and considerate of others. Keep listening to your clients/patients and continue to give them life-changing solutions—including neurofeedback!

In their workshop "Environment Toxins and Electromagnetic Radiation" at the ISNR Annual Conference August 2023 in Dallas, Texas; David Cantor and Robert Turner conducted a two-part workshop. "Part I: Toxins: The Roots of All Evils Impact on Brain Function and Behavior," and "Part II: Environmental Toxins: Electromagnetic Radiation."

Toxins are also a significant health problem. David Cantor notes: *Neurotoxins (nerve poisons) are an extensive class of exogenous chemical or electromagnetic neurological insults which can adversely affect function in both developing and mature nervous tissue. The term can also be used to classify endogenous compounds which when abnormally concentrated can prove neurologically toxic. Though neurotoxins are often neurologically destructive, their ability to specifically target neural components is important in the study of nervous systems. Common examples of chemo-neurotoxins include lead, ethanol, glutamate, nitric oxide (NO), botulinum toxin, tetanus toxin, and tetrodotoxin. Electromagnetic toxins are now far more pervasive and equally destructive.*

Cantor also reports, *Neurotoxin activity can be characterized by the ability to inhibit neuron control over ion concentrations across the cell membrane, or communication between neurons across a synapse. Local pathology of neurotoxin exposure often includes neuron excitotoxicity or apoptosis but can also include glial cell damage. Macroscopic manifestations of neurotoxin exposure can include widespread central nervous system damage such as mental retardation, persistent memory impairments, epilepsy, and dementia. Additionally, neurotoxin-mediated peripheral nervous system damage such as neuropathy or myopathy is common. Support has been shown for a number of treatments aimed at attenuating neurotoxin-mediated injury, such as antioxidant, antitoxin and ethanol administration.*

Cantor notes, *Toxic agent-induced immunostimulation can cause autoimmune diseases, in which healthy tissue is attacked by an immune system that fails to differentiate self-antigens from foreign antigens. For example, the pesticide dieldrin induces an autoimmune response against red blood cells, resulting in hemolytic anemia. Toxic chemicals like BPA (biphenol A) and phthalates that can harm the immune system. They can be found in: Plastics labeled with the #3 or #7 recycling mark. Food and drink can linings.*

We live in a world that is rapidly advancing the use of electronics and portable devices. Smart phones connected to specialized wrist watches, and a variety of electronic devices are commonplace in our homes, cars, and travel accessories. Many of these are devices used to monitor and/or improve our health.

Unfortunately, many of these electronics and devices

emit EMR (electromagnetic radiation—the emission of energy as electromagnetic waves). They may also emit radio frequency (RF) radiation which is a type of non-ionizing radiation. It is used in many broadcast and communication applications, including cellular phones and wearable technologies. RF radiation is the transfer of energy by radio waves and lies in the frequency range between 3 kilohertz (kHz) to 300 gigahertz (GHz)1. RF radiation has insufficient energy to break chemical bonds or remove electrons (Australian Government, 2024).

EMR and RFR both have a direct impact on brain function, behavior and our overall health. Turner (2023) reports that:

1) Artificial EMR affects the nervous system.

2) EMR contributes (causative vs. correlative) to many symptoms of mental ill-health and neurological dysfunction (sleep, anxiety, OCD, fatigue, HA, etc.).

3) Reducing exposure to EMR will reduce/eliminate

associated symptoms, improve function, and decrease inflammation.

4) EMR interferes with responses to NFB and other neuromodulation modalities (train vs. "treat"). Turner (2023) notes that the most unrecognized environmental toxin, increasing at exponential rates, is human-derived ("artificial") electromagnetic radiation (EMR) in the forms of advancing technology: wireless connectomics and more powerful pervasive telecommunications. The preponderance of the research found that exposure to RFR or ELF EMF produces oxidative effects of free radicals and damages DNA (Lai, 2023).

The following chart (Burrell, 2024) indicates the various health risks associated with electromagnetic fields.

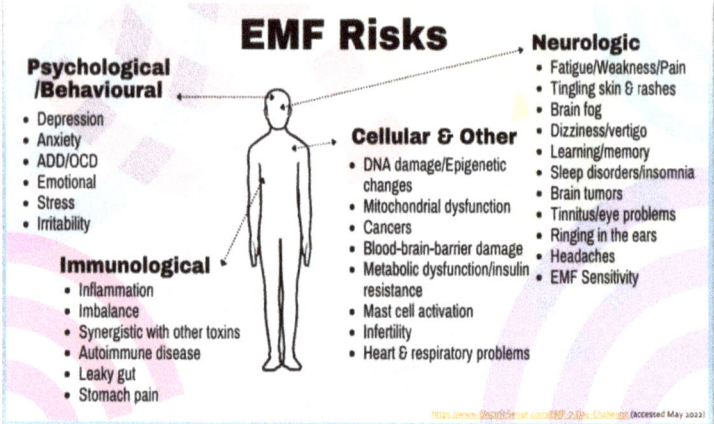

**EMF Risks**

**Psychological /Behavioural**
- Depression
- Anxiety
- ADD/OCD
- Emotional
- Stress
- Irritability

**Immunological**
- Inflammation
- Imbalance
- Synergistic with other toxins
- Autoimmune disease
- Leaky gut
- Stomach pain

**Cellular & Other**
- DNA damage/Epigenetic changes
- Mitochondrial dysfunction
- Cancers
- Blood-brain-barrier damage
- Metabolic dysfunction/insulin resistance
- Mast cell activation
- Infertility
- Heart & respiratory problems

**Neurologic**
- Fatigue/Weakness/Pain
- Tingling skin & rashes
- Brain fog
- Dizziness/vertigo
- Learning/memory
- Sleep disorders/insomnia
- Brain tumors
- Tinnitus/eye problems
- Ringing in the ears
- Headaches
- EMF Sensitivity

Cantor and Turner (2023) note the effects of EMR on EEG:

**The "BASICs" of EMR effects on EEG:**

**B – Beta**
(elevated fast 18-40 Hz)

**A – Alpha**
(PDR slowing; diminishing power/amplitude)

**S – Slowing**
(focal slowing, maximal left>>right temporal/frontal)

**I – Isolated Epileptiform Discharges**
(birth and evolution of IEDs)

**Cs – Circadian (Sleep)**
(rapid drowsiness/sleep; poor EO/EC differentiation; persistent PSWY)

# CHAPTER FOUR

## Formulating a Functional Neurofeedback Treatment Plan

The majority of health care professionals need to develop a treatment plan for the clients/patients to whom they provide services. Neurofeedback clinicians are no different. After an initial consultation or meeting, a comprehensive history needs to be taken. Since neurofeedback clinicians come from various professional disciplines, treatment planning may differ from one discipline to another.

In this chapter we simply want to outline the various areas that may and/or should be considered as an individualized treatment plan is formulated. Functional neurofeedback is a holistic approach to health care. The body and mind work together.

Treatment planning areas to be addressed include but are not limited to the following:

**Reason for Contact:** Presenting problem(s), when symptoms began, severity of symptoms, current and/or previous interventions.

**Personal History:** Current status, marriage/family, children, sexuality, work, school, hobbies, interests, social history, lifestyle, education, exercise and activity, cultural aspects, support systems, home and/or environmental risk factors, abuse neglect, trauma, etc.

**Physical Health:** History of any current health problems, medications, procedures, labs, tests, X-rays, past medical history, surgical history, allergies, smoking, alcohol use, recreational drug use, diet, nutrition, supplements, previous ER visits, etc.

**Family Health History:** Parents siblings, children, abuse, neglect, etc.

**Mental Health History:** Current or previous diagnosis, cognitive deficits/functioning, appearance/attire/ presentation, strengths, weaknesses, judgement, client/ patient/family concerns, behavioral concerns, mood, negative feelings, affect, personal insight, how many times in therapy, how many different therapists, etc.

**Head Injury:** One of the most common areas not asked of clients/patients by health care professionals is history of head injury. This is important for neurofeedback practitioners to know as head injury brain wave patterns often show up in QEEG brain maps and impact a variety of physical and mental functions.

The items and areas listed above are not an all-inclusive list, but rather indicative of information that neurofeedback clinicians aim to acquire in order to prepare a comprehensive treatment plan. Why? A comprehensive history gives the clinician insight into the etiology of physical and mental health problems. Functional medicine investigates the etiology and cause(s) of health problems. Functional neurofeedback also seeks to find the etiology of problems being presented to the clinician by considering all the information above that may go into conducting a comprehensive assessment.

Good history provides the clinician with more tools to help formulate a comprehensive treatment plan that takes into account lifestyle changes. As noted in Chapter 3, Dr. Robert Turner, addresses lifestyle factors in assessment of patients and their overall general health and emotional functioning.

## Obstacles That May Affect Neurofeedback Benefits and Outcomes

**Diet:** Processed foods and fast foods are examples of foods that may not provide the client/patient with proper nutrition and may have additives that are unhealthy. Some foods have additives that may be toxic.

**Sleep:** Sleep is essential for proper health and to replenish our bodies. Poor sleep or not enough sleep effects our cognitive functioning.

**Exercise:** Exercise is important for overall health. It is important for maintaining physical health and brain health.

**Physical health:** One's overall physical health effects brain function. Clients/patients with severe physical health problems may take longer to respond to neurofeedback and in worse case scenarios, may not respond at all to neurofeedback interventions.

**Medications:** Psychotropic medications effect brain waves and often have undesirable side effects. Clients/patients taking psychotropic medications must work with their prescriber and notify them that they are participating in neurofeedback with the intent to reducing or eliminating those medication over time.

**Substance abuse:** Clients/patients can become addicted to both prescription medications as well as recreation drugs.

**Recreational drugs:** Like prescription drugs, clients/patients using recreational drug use including alcohol and cannabis may not benefit from neurofeedback if they are not willing to reduce and ultimately eliminate their use during the time they are participating in neurofeedback.

**Social relationships:** Stressful social relationships (i.e., domestic violence, abuse, consistent fighting, dislike of work, kids being bullied at school can cause the client/patient to not obtain significant benefit from neurofeedback.

### Protocol Planning

We would caution the reader to be aware of the fact that there is not one single best protocol to use when targeting a specific symptom, i.e., anxiety, depression, attention deficit, poor sleep, etc. Using a QEEG is best for determining protocols and locations based upon the client/patients' presenting symptoms.

The number of neurofeedback sessions is also not a fixed limit or amount. Historically, the field has noted 30–40 sessions as a standard intervention with neurofeedback. In recent years, ISNR has revised that range to be from 20–40 sessions. However, based upon some of the factors listed above as well as certain conditions 50 or more sessions may be necessary.

### Adverse Effects

A comprehensive assessment can give the clinician an advantage of foreseeing possible adverse effects. For example, some people who are very depressed or anxious, or who have other types of mental health problems, can find that relaxation doesn't help. It might even make them feel worse. Clients/patients with histories of trauma and/ or PTSD may find relaxation techniques challenging.

Rogel et al. (2015) reported adverse effects ranging from emotional effects such as anxiety, agitation, and emotional liability to physiological ones such as enuresis, muscle twitches and ticks.

Adverse side effects can be transient or long term. They can be mild (e.g. headaches, fatigue, irritability, or being spacey) or severe (e.g. seizures, depression, manic attacks, and memory problems). Many of these mild side effects pass within a short time after the session.

In their study, Rogel et al. (2015) reported the following list of adverse effects: ringing in ears, stomach bloating, stomach cramps, diarrhea, nausea, shortness of breath, runny nose, vertigo, headaches, sore eyes, blurred vision, fatigue, restlessness, increased dreaming/nightmares, trouble sleeping, mood swings, memory problems, confusion, trouble concentrating, feeling detached, tearfulness, anxiety attacks, anger, moodiness, irritability, feeling high, and nervousness.

**Medication Sensitivity**

Some individuals have sensitivity to medications and do better with smaller doses or have more exaggerated side effects. Those who are sensitive to medications often are sensitive to neurofeedback and do better with shorter sessions, i.e., 15–20 minutes vs. a standard 30-minute session.

**Medication Reduction and Withdrawal Symptoms**

When clients/patients are taking psychotropic medications, they need to work with both the neurofeedback clinician and medication prescriber to reduce dosage over time and in many cases discontinue the medication(s). The typical time between dosage reductions is 1–2 weeks; however, as neurofeedback providers we encourage a 3-week period to allow time for the neurofeedback to counter medication side effects and withdrawal symptoms.

In Anatomy of an Epidemic, Robert Whitaker (2010, "Chapter 5: The Withdrawal Syndrome") discusses the withdrawal syndrome that can occur when people try to stop taking psychiatric medications. He explains how the withdrawal syndrome can be severe and long-lasting, and how it can lead to relapse and re-hospitalization.

Discontinuing an antidepressant medication usually involves reducing your dose in increments, allowing 2–6 weeks or longer between dose reductions.

Patients/clients with a history of trauma and/or abuse often have difficulty disclosing these life experiences. This is especially true of those who experienced sexual abuse as a child. Often this history is kept buried and silent due to embarrassment until a clinician specifically asks the question.

**Progress Tracking**

Neurofeedback is an incredibly powerful and beneficial intervention. Clients/patients will have varied responses to the first few sessions and the entire process based upon their history and background. Many will notice a benefit within the first session or two. Others may need 5–10 sessions or more to detect benefits based upon their individual history, diagnosis, and health.

Progress tracking is important in order to determine the efficacy and benefits of specific NFB protocols as the clinician addresses the various problems/symptoms the client/patient is wanting to address.

We encourage the clinician to work with the client/patient to create a list of problems they want to address. A list of 10–12 areas is sufficient to get started. As problems or symptoms resolve, new or additional symptoms can be addressed.

We encourage our clients/patients to complete progress trackers the day before or on the day of each session. In some instances, we request they fill out a progress tracker daily in order to monitor progress/benefits for a specific disorder, i.e., migraine headaches, poor sleep, anxiety/ panic attacks.

**Discharge Planning**

Discharge planning begins the day the client/patient signs up for NFB and should be individualized. Are there follow-up contacts with other health care providers? Are there additional interventions for the client/patient to consider? Does he/she need to be referred to another health care provider for any reason? What should the client/patient do if symptoms recur?

Winding down NFB can be done in different ways. We encourage clinicians to get the client/patient down to one session per week and after five sessions assess if the benefits are consistent between sessions and there is no slippage or relapse of symptoms. Then the trainee is scheduled to come back for a single session in 2 weeks. That is repeated again after 2 more weeks, and then the trainee is instructed to come back in 3 weeks. If there is no slippage/recurrence of symptoms the client/patient is discharged from NFB.

Longo and Bingham (2022) report: *A summary report is both a professional courtesy and another way to let those working in the medical field in your area know that you offer neurofeedback. We ask the client during the intake process to provide the name of their primary care, referring physician if different, therapist as applicable, and offer to send them a copy of the report. Providers gain a better understanding of the neurofeedback process and measurable improvement, which in turn helps their modalities have a more significant impact. Parents find this information helps them advocate for their child within the school setting and IEP meetings. The report includes intake information and QEEG results. We send out a report at the beginning of training and every 15 sessions after that with a progress summary and highlights from the most recent QEEG results.*

**Post NFB**

As noted above, there is no general rule for determining when a client has successfully completed neurofeedback. For most mental health disorders such as anxiety, depression, etc., the conventional wisdom has dictated 30–40 sessions. With advances in equipment and software and increased knowledge in running sessions, ISNR now indicates a range of 20–40 sessions. For persons with severe disorders such as a traumatic brain injury, the number of sessions can reach into the 50s or 60s—or more.

The majority of clients/patients we work with, who have finished with NFB, have long lasting benefits. However, as part of discharge planning, we always advise them that life events and circumstances occur and symptomology returns (i.e., a stressful life event that causes severe anxiety), coming in for a "tune-up" is always an option. We describe a tune-up as generally coming back into the office for another 3–5 sessions to get brain function improved.

# CHAPTER FIVE

## Using Function Neurofeedback in your Practice

*In their book, Beginning Neurofeedback in Your Practice: A Guide for Clinicians Using Neurofeedback From Intake to Discharge (Longo & Bingham, 2022), Longo and Bingham state: We encourage you to join organizations that have a focus on neurofeedback and offer workshops on adding neurofeedback to your practice, marketing, strategies, etc., such as the International Society for Neuroregulation and Research (ISNR), the Association for Applied Psychophysiology and Biofeedback (AAPB), and one of the many regional societies that often hold annual conferences addressing issues related to neurofeedback and biofeedback. We also encourage you to become Board Certified in Neurofeedback. There are a few organizations that provide this certification; however, we strongly support and recommend the Biofeedback Certification International Alliance (BCIA).*

Using functional neurofeedback in your practice is not a challenging or overwhelming concept. It is simply a matter of applying the principles discussed in previous chapters. One significant change that most practitioners make is how they conduct an initial consultation with a perspective new client/patient.

Taking a detailed history is a foundational aspect of functional neurofeedback. In this instance, the clinician has two basic options: 1) to create intake forms that the clinician or another clinical staff-person fills out during the initial intake, or 2) create forms that the client/patient fills out after an initial consult (hard copy or online).

Neurofeedback is a powerful tool. It can be especially

valuable when clients have exhausted other traditional treatments without positive effect. As neurofeedback teaches the client's brain to function more effectively, other therapies are often augmented for additional benefit.

How you market your practice is also important.

Longo and Bingham (2022) note, *before you dive into branding and marketing, you need to consider the audience you'll serve, the disorders you'll specialize in, and the scope of your practice based upon your degree and licensure.*

They suggest, *Who is in your community or within driving distance? Working professionals? Mothers of young children? Parents of teens? Grandparents? For example, one practice in a large city is well known for nationally ranked elementary and secondary schools. The community makeup is very heavily weighted towards families, enabling them to focus heavily on children and, to a lesser extent, on younger mothers. The practice is in a bedroom community outside of a large city but has not chosen to focus on technology or other working professionals in part because the commutes are so long, preventing them from attending sessions except for the very late evening. If you're not already intimately familiar with your potential customers' demographic makeup, speak with your chamber of commerce or your local Service Corps of Retired Executives (SCORE) to help refine our target audience.*

The majority of the practitioners we know using a functional neurofeedback model do not conduct an initial QEEG brain map or begin neurofeedback training before completing all intake paperwork. This is important because comprehensive physical and mental health histories can help validate findings in the QEEG report as well as determine if there are any possible barriers to neurofeedback training (i.e., excessive physical health problems, psychotropic medications that may impact specific brain waves, among other challenges).

A functional neurofeedback model would also include a well-established referral source. It is not uncommon to have clients/patients who have been seeing health care providers who are not familiar with neurofeedback. In some cases, especially when the prescriber does not know about neurofeedback and its benefits, medication reduction as a treatment goal could be compromised. As mentioned on page 39, we encourage clients/patients to do reductions once every 3 weeks to allow time for the neurofeedback to counter medication side effects and withdrawal symptoms.

Therefore, we recommend developing a system of referral sources of physicians and health care practitioners who will work with you and your practice and make mutual referrals. Longo and Bingham (2022) suggest: It is important to ask for referrals from your network of friends, providers, associates, and especially clients. Typically, these referred individuals are, to some extent, pre-sold. Rather than needing the seven-plus contacts with the idea of neurofeedback, they are ready to sign up with one phone call and perhaps a visit to your office.

*Local providers are more likely to refer to you if they are educated about what neurofeedback can do and how it complements their business. As you build relationships through networking, community events, and other interactions, local providers can see how combining neurofeedback with what they offer a client increases overall improvement and healing.*

# CHAPTER SIX

## Challenges of Functional Neurofeedback

Neurofeedback is a highly specialized intervention that requires targeted training, mentoring, equipment, software, and commitment to continuous learning. It is an amazing tool that can have a profoundly positive impact on the brain. The improvement clients can experience as their negative symptoms decrease or are eliminated has tremendous positive implications for their lives, the lives of their family, and society as a whole. In our opinion, it is well worth the effort. We are excited to introduce the concept of functional neurofeedback into the field, both for our own growth in an area that will constantly challenge us and for the benefit of the clients whose lives we will change for the better.

However, functional neurofeedback is no different than any other method of neurofeedback when it comes to challenges clinicians might face.

Client/patient compliance is one of the challenges NFB practitioners face. Healthcare and client/patient understanding of disorders, interventions and self-care is often a significant obstacle. Our culture and society have become accustomed to quick fixes. Take a pill... feel better. Have a procedure done... problem resolved. Have surgery, problem repaired or removed.

Our expectation of client/patient compliance becomes more refined with FNFB. If there are significant health related issues. Some issues can interfere with the benefits of NFB. Other medical conditions would require intervention from a physician before starting neurofeedback. Below is a partial list of symptoms we encourage clinicians to monitor.

Sleep
Teeth grinding
Difficulty falling asleep
Bedwetting
Difficulty staying sleep
Periodic leg movements
Difficulty waking up
Restless leg
Dysregulated sleep cycle
Restless sleep
Narcolepsy
Sleep apnea
Night sweats
Sleep walking
Night terrors
Snoring
Nightmares or vivid dreams
Talking during sleep

Concentration
Difficulty completing tasks
Not listening
Difficulty following directions
Poor concentration
Difficulty making decisions
Poor drawing ability
Difficulty organizing personal
time or space
Poor math

Difficulty remembering names
Poor short-term memory
Difficulty shifting attention
Poor sustained attention
Difficulty shifting tasks
Poor verbal expression
Difficulty thinking clearly
Poor vocabulary
Difficulty understanding
conversations
Poor word finding
Distractibility
Reading difficulty
Lack of alertness
Slow thinking
Lacking common sense
Unmotivated
Messy handwriting

Sensory
Auditory hypersensitivity
Tinnitus
Chemical sensitivities
Vertigo
Motion sickness
Visual deficits
Poor body awareness
Visual hypersensitivity
Somatosensory deficits

## Behavior
Addictive behaviors
Lack of sense of humor
Aggressive behavior
Lack of social interest
Anorexia
Manipulative behavior
Autistic stimming
Motor or vocal tics
Binging and purging
Nail biting
Class clown
Oppositional or defiant behavior
Compulsive behaviors
Poor eye contact
Compulsive eating
Poor grooming
Crying
Poor social or emotional reciprocity
Excessive talking
Poor Speech articulation
Hyperactivity
Rages
Impulsivity
Self-injurious behavior
Inflexibility
Stuttering
Lack of appetite awareness
Trouble doing anything because felt bad

## Emotion
Agitation
Anger
Lack of emotional awareness
Anxiety
Lack of pleasure
Depression
Lack of social awareness
Difficult to soothe
Low self-esteem
Dissociative episodes
Mania
Easily embarrassed
Mood swings
Emotional reactivity
Obsessive negative thoughts
Fears
Obsessive worries
Feelings of unreality
Panic attacks
Flashbacks of trauma
Paranoia
Impatience
Suicidal thoughts
Phobias
Sexual indifference
Worry
Victim mentality
Socially inappropriate
Socially cavalier
Self-deprecation
Passive aggressiveness
Over control of emotion
Irritability
Hyperactive attention
Hyper vigilance
Hyper arousal
Excessive self-concern
Excessive rationalization
Emotionally impulsive
Emotional rumination
Dislike of novelty.

### Cognitive
Attention problems
Auditory tone processing problems
Auditory verbal sequence Problems
Categorization problems
Decision making problems
Declarative & episodic memory problems
Digit span problems
Event sequence Problems
Math problems (acalcula)
Motivation problems
Poor dialogue organization
Poor facial recognition
Poor figure memory
Problem solving difficulties
Procedural memory Problems
Reading comprehension
Short term memory difficulty
Short term verbal memory problems
Short term visual memory problems
Spatial sequencing problems
Tone sequence problems
Verbal sequencing problems
Working memory problems

### Physical
Allergies
Nausea
Asthma
PMS symptoms
Chronic constipation
Poor balance
Clumsiness
Poor fine motor coordination
Difficulty walking or moving
Poor gross motor coordination
Difficulty working
Reflux
Effort fatigue
Rigidity
Encopresis
Seizures
Fatigue
Skin rashes
Heart palpitations
Spasticity
High blood pressure
Stress incontinence
Hot flashes
Sugar craving and reactivity
Immune deficiency
Sweating
Irritable bowel
Tachicardia
Low muscle tone
Tremor
Muscle tension
Urge incontinence
Muscle twitches
Abdominal bloating
Always sickly
Insomnia
Amnesia
Anxiety attacks
Labored breathing
Aphonia (loss of voice above a whisper)
Lump in throat
Menstrual irregularity
Bulimia

**Physical (con't)**
Paralysis
Ringing in ears
Dizziness
Spasms
Sudden weight fluctuation
Excessive menstrual bleeding
Unconsciousness
Urinary retention
Fainting spells
Visual blurring
Vomiting
Fits or convulsions
Food intolerances
Weakness
Frigidity (absence of orgasm)
Weight loss
Indigestion
Heartburn

**Pain**

Abdominal pain

Muscle pain
Chronic aching pain
Muscle tension headaches
Chronic nerve pain
Sciatica
Fibromyalgia pain
Sinus headaches
Jaw pain
Stomach aches
Joint pain
Trigeminal neuralgia
Headaches
Burning pains in rectum, vagina, or mouth
Extremity pain
Other bodily pains
Chest pains
Dysmenorrhea (painful menstruation)
Dysmenorrhea-other
Dyspareunia (painful sexual intercourse)
Dysuria (painful urination)

Some of these conditions and metabolic categories would require lab work and other testing before we would begin a client/patient on neurofeedback.

Many perspective clients/patients come to the office with a host of psychotropic medications. It is not unusual to hear the perspective trainee tell the clinician that they have been taking these medications for years and symptomology is still present. In these cases, medication reduction or elimination should be discussed with the trainee prior to beginning NFB.

**Lifestyle Challenges**

Robert Turner's M.E.D.S. Model (Chapter 3) can be

challenging for many clients/patients are their families to adapt.

**Sleep Hygiene:** Many people have poor sleep hygiene and bedtime habits. Bedrooms should be set up with sleep in mind. The room should be dark and quiet. Use of computers, TVs, phones, and tablets should be discontinued at least one hour before bedtime. Bedtime and wake up time should be consistent allowing for 7–9 hours of sleep.

**Electronics:** As noted above electronics should not be used one hour before bedtime. Cell phones, computers, TVs, tablets and all electronic flat screen devices emit blue light triggering the brain to think it is daytime.

**Exercise:** Humans are active and exercise is important for all age groups. We have become a society that has gradually shifted towards a more sedentary lifestyle. Working at a desk, working on computers, and sitting is not good for our health. It is important to make sure clients/patients engage in some form of physical activity each day. Even the simple act of walking provides us with some of the exercise we need each day.

**Diet and nutrition:** We are what we eat. Our country has made shifts towards fast and processed foods which are advertised on the radio, TV, and in magazines and newspapers. Proper nutrition is essential for overall physical health and brain health.

We have found over the course of time and working with hundreds of patients that three significant life experiences can work against the trainee receiving the best benefits from neurofeedback. Excessive use of psychotropic medications, excessive physical health problems, and problematic/stressful social relationships can all slow down the benefits of neurofeedback.

Problematic social relations may include situations such as a child living in a domestic violence home, a child being bullied at school, sibling abuse.

For adults, social situations such as a problematic boss at work, domestic violence, marital discord, and problematic relationships with family or friends can all be stressful. Often these situations are 24/7 stressors in the trainee's life.

**Training Consistency:** In an ideal setting, clients/patients would come into the office for a neurofeedback session twice per week (minimally once per week). Neurofeedback is cumulative. Inconsistent training and missed sessions will slow down the trainee's overall progress.

Home/remote training (the client/patient is provided software and equipment to conduct NFB sessions on themselves or a family member in their own home); can also present challenges. Proper hookup and running of sessions is important. Often the client/patient will not run routine sessions or sessions are run late in the evening. Clients/patients need to be properly trained not only in running a session but on the importance of following a routine session schedule at home as if they were having sessions in the office.

The above outlined situations can all present challenges to the clinician providing neurofeedback services; the majority of which can be resolved with proper guidance and review of services being provided.

# References

Allsopp, K., Read, J., Corcoran, R., & Kinderman, P. (2019). Heterogeneity in psychiatric diagnostic classification. Psychiatry Research, 279, 15–22. https://doi.org/10.1016/j.psychres.2019.07.005

Australian Government (2024). Radiofrequency radiation. https://www.arpansa.gov.au/understanding-radiation/what-is-radiation/non-ionising-radiation/radiofrequency-radiation

Black M. M. (2008). Effects of vitamin B12 and folate deficiency on brain development in children. Food and Nutrition Bulletin, 29(2 Suppl), S126–S131. https://doi.org/10.1177/15648265080292S117

Bland, J. (2017). Defining function in the functional medicine model. Integrative Medicine (Encinitas, Calif.), 16(1), 22–25.

Bland J. S. (2018). The natural roots of functional medicine. Integrative Medicine (Encinitas, Calif.), 17(1), 12–17.

Bland J. S. (2019a). What is evidence-based functional medicine in the 21st century? Integrative Medicine (Encinitas, Calif.), 18(3), 14–18.

Bland J. S. (2019b). Systems biology meets functional medicine. Integrative Medicine (Encinitas, Calif.), 18(5), 14–18.

Burrell, L. (2024). The invisible danger lurking in your home. https://www.electricsense.com/emf-7-day-challenge/

Cantor, D. & Turner, R. (2023). Toxins: The roots of

all evils—Impact on brain function and behavior. Presented at ISNR 2023 Conference August 26, Dallas, TX.

Carabotti, M., Scirocco, A., Maselli, M. A., & Severi, C. (2015). The gut-brain axis: Interactions between enteric microbiota, central and enteric nervous systems. Annals of Gastroenterology, 28(2), 203–209.

Chang, S., & Lee, T. H. (2018). Beyond evidence-based medicine. The New England Journal Of Medicine, 379(21), 1983–1985. https://doi.org/10.1056/NEJMp1806984

Coben, R., Hammond, D. C., & Arns, M. (2019). 19 Channel Z-Score and LORETA neurofeedback: Does the evidence support the hype? Applied Psychophysiology and Biofeedback, 44(1), 1–8. https://doi.org/10.1007/s10484-018-9420-6

Dholakia, K. R., Dharmarajan, T. S., Yadav, D., Oiseth, S., Norkus, E. P., & Pitchumoni, C. S. (2005). Vitamin B12 deficiency and gastric histopathology in older patients. World Journal of Gastroenterology, 11(45), 7078–7083. https://doi.org/10.3748/wjg.v11.i45.7078

Fisher, S. (2022). Webinar: Traumatized brain: Recent neuroscience findings and their implications presented by Sebern Fisher, MA, BCA. https://isnr.org/events/traumatized-brain-recent-neuroscience-findings-and-their-implications-presented-by-sebern-fisher-ma-bca

France, N. (2022). How not to use brain scans in neuroscience. https://neurosciencenews.com/neuroimaging-sample-21226/

Healy, D. (2004). Let them eat Prozac. New York University Press, New York.

International Society for Neuroregulation and Research (2024, June 4). https://isnr.org .

Krepel, N., Egtberts, T., Sack, A. T., Heinrich, H., Ryan, M., & Arns, M. (2020). A multicenter effectiveness trial of QEEG-informed neurofeedback in ADHD: Replication and treatment prediction. NeuroImage: Clinical, 28, 102399.

Kuhn, Thomas S. (1970). The structure of scientific revolutions. Chicago: The University of Chicago Press.

Lai, H. (2023). Effects of radio frequency radiation exposure on free radical-related cellular processes (290 studies). https://www.saferemr.com/2018/02/effects-of-exposure-to-electromagnetic.html

Lane, C. (2013). The NIMH withdraws support for DSM-5. https://www.psychologytoday.com/us/blog/side-effects/201305/the-nimh-withdraws-support-dsm-5. Retrieved June 4, 2024

Longo, R. E. & Bingham, B. (2022). Beginning neurofeedback in your practice: A guide for clinicians using neurofeedback from intake to discharge. Greenville, SC: FNNR.

Longo, R. & Sherlin, L. (2024). Scope of practice & ethics in neuromodulation (parts 1 & 2). Part 1 https://www.youtube.com/watch?v=Y-qhRX6iBZo&t=1523s and part 2 https://www.youtube.com/watch?v=2QMKbB_2fyc

Núñez-Jaramillo, L., Herrera-Solís, A., & Herrera-

Morales, W. V. (2021). ADHD: Reviewing the causes and evaluating solutions. Journal of Personalized Medicine, 11(3), 166. https://doi.org/10.3390/jpm11030166

Raskin, J. (2012). What is the DSM-5 definition of a mental disorder? https://www.saybrook.edu/unbound/defining-mental-disorders-dsm-5-style

Robbins, J. (1998). Wired for miracles. Psychology Today, 31(3), 40–47.

Robbins, J. (2000). A symphony in the brain. New York. Grove Hills.

Rogel, A., Guez, J., Getter, N., Keha, E., Cohen, T., Amor, T., & Todder, D. (2015). Transient adverse side effects during neurofeedback training: A randomized, sham-controlled, double-blind study. Applied Psychophysiology and Biofeedback, 40(3), 209–218. https://doi.org/10.1007/s10484-015-9289-6

Science Daily (2019, July 8). Psychiatric diagnosis "scientifically meaningless." Science Daily. https://www.sciencedaily.com/releases/2019/07/190708131152.htm

Shonkoff, J. P., Garner, A. S., Committee on Psychosocial Aspects of Child and Family Health, Committee on Early Childhood, Adoption, and Dependent Care, & Section on Developmental and Behavioral Pediatrics (2012). The lifelong effects of early childhood adversity and toxic stress. Pediatrics, 129(1), e232–e246. https://doi.org/10.1542/peds.2011–2663

Soutar, R. (2021). Holistic neurointegration: The NewMind model. A bio-psycho-social qEEG guided neurofeedback method. NewMind, Woodstock, GA.

Swatzyna, R. J. (2009). The elusive nature of mild traumatic brain injury. Biofeedback 37(3), 92–95.

The Institute for Functional Medicine (2024, June 4). https://www.ifm.org/functional-medicine.

The Institute for Functional Medicine (2022, August 7). Learn proven techniques to help patients make lifestyle changes. https://tinyurl.com/2exhevf4

Turner, R. (2023). Workshop: Environmental toxins: Electromagnetic radiation (EMR) impact on brain function and behavior. Presented at the ISNR 2023 Conference August 26, 2023. Dallas, TX.

Van der Kolk, B. (2014). The body keeps the score: Brain, mind and body in the healing of trauma. Viking, New York.

Wexler, A., Nagappan, A., Kopyto, D., & Choi, R. (2020). Neuroenhancement for sale: Assessing the website claims of neurofeedback providers in the USA. Journal of Cognitive Enhancement, 4, 379–388.

Whitaker, R. (2010). Anatomy of an epidemic: Magic bullets, psychiatric drugs, and the astonishing rise of mental illness in America. Crown, NY.

Winerman, L. (2013). NIMH funding to shift away from DSM categories. Monitor on Psychology, 44(7).